Memorable Japanese Motorcycles
1959-1996

Doug Mitchel

4880 Lower Valley Road, Atglen, PA 19310

Acknowledgments

I want to take a moment to thank all of my friends that ride these great machines. Without their time and efforts, I would be forced to seek employment elsewhere. It would be impossible to repay them for their time spent with me, so I offer this brief moment in the sun to them as a way of showing my appreciation. Thanks gang.

Within the pages of this book, we have attempted to show a healthy cross section of the machines that have filled our lives for the last 40 years. As usual, there are many more we would love to see. Please don't feel slighted if a favorite of yours was omitted. Time and logistics limit me to use what I can find, and I know there are plenty of great machines waiting to be showcased. Feel free to contact me if you have a cycle you feel belongs in the pages of another volume.

As always, ride free, and keep the rubber side down.

Copyright © 1997 by Doug Mitchel
Library of Congress Catalog Number: 97-65625

All rights reserved. No part of this work may be reproduced or used in any form or by any means--graphic, electronic, or mechanical, including photocopying or information storage and retrieval systems--without written permission from the copyright holder.

Layout by Blair R.C. Loughrey

ISBN: 0-7643-0235-3
Printed in Hong Kong

Published by Schiffer Publishing Ltd.
4880 Lower Valley Road
Atglen, PA 19310
Phone: (610) 593-1777; Fax: (610) 593-2002
E-mail: Schiffer@aol.com
Please write for a free catalog.
This book may be purchased from the publisher.
Please include $2.95 for shipping.
Try your bookstore first.

We are interested in hearing from authors with book ideas on related subjects.

Table of Contents

Introduction 4

Japanese Motorcycles — 1959-1996 5

1959 Lilac 6
1965 Kawasaki B8-T 10
1965 Marusho S.T.500 12
1966 Honda CB-77 Super Hawk 15
1966 Honda Sport 160 19
1966 Kawasaki W-1 22
1966 Suzuki T10 25
1966 Yamaha Big Bear Scrambler 29
1967 Honda CB-450 Black Bomber 32
1967 Honda CL-77 Scrambler 36
1967 Honda Dream 39
1967 Honda Super 90 43
1967 Suzuki X-6 Hustler 46
1969 Honda CB750 Sandcast 49
1971 Honda CL-350 52
1971 Yamaha R5 350 55
1973 Honda CB350 58
1973 Kawasaki H-2 61
1973 Kawasaki Z-1 64
1974 Honda CB200 68
1974 Kawasaki H-1 71
1974 Yamaha RD 350 73
1975 Honda CB400/4 76

1976 Honda GL1000 Limited Edition 79
1978 Kawasaki Z1-R 82
1978 Kawasaki Z1-R TC 85
1978 Yamaha XS-750 88
1981 Honda CM400A 91
1981 Yamaha 550 Seca 94
1982 Honda CBX 97
1982 Kawasaki KZ1000R 100
1983 Honda CX650T 103
1983 Kawasaki GPz550 106
1983 Suzuki Katana 1100 109
1983 Yamaha 900 Seca 112
1984 Honda Nighthawk S 115
1984 Kawasaki 900 Ninja 118
1985 Kawasaki Eliminator 121
1985 Yamaha V Max 124
1986 Honda V65 Magna 127
1986 Suzuki GSX-R 750 130
1986 Suzuki RG500 Gamma 133
1989 Honda GB500 Tourist Trophy 136
1990 Honda RC-30 139
1996 Honda CBR600 F3 143
1996 Honda VFR 750 146

Introduction

Motorcycles have played a part in America's history since the turn of the twentieth century. Early machines were little more than bicycles powered by internal combustion engines, but soon began growing into useful pieces of machinery.

By the middle of the 1950s, America found itself with only one remaining manufacturer of two wheeled, motorized machines. The British had begun importing their brands to our shores, and the competition was beginning to heat up. As if things weren't confusing enough, a little country named Japan soon started making inroads into the U.S. market.

The first machines from Japan were simple, lightweight units that could be ridden by anyone, and opened up new roads for the neophyte rider. As the years progressed, the offerings from the Japanese makers became the ones to beat.

By the end of the 1960s, the British had pulled their cycles from our market due to their inability to compete with the selections available at the dealerships selling Japanese equipment. By 1983, even the giant Harley-Davidson was forced to turn to the government for protection from the Japanese.

Japanese manufacturers continue to fill our garages with well made, efficient models that range from basic twin cylinder units all the way up to six cylinder touring bikes that seem like two wheeled automobiles. U.S. manufacturers have taken to doing things in the style of the Asian builders, and we have better machines to show for it. We now have the ability to choose a motorcycle to fit our needs, no matter what they are. Odds are, if it were not for the invasion of inexpensive Japanese cycles in the U.S., we might have been forced to remain trapped in our four wheeled transportation forever.

Japanese Motorcycles — 1959-1996

1959 Lilac

Owner: The Jim Kersting Family Collection

Almost in lock step with the Honda Motor Company, the Lilac first appeared on these shores as a 1959 model. Equipped with a 246cc, transverse V-twin motor, and shaft drive, the Lilac was rather luxurious when compared to the competition's offerings.

As a rule, the machines from Lilac were all rather expensive, and well equipped. Not only did they build V-twin powered cycles, but also sold single cylindered units, and tiny 49cc mopeds. Similar to the motorcycle manufacturers that proliferated in the U.S., Lilac was soon to fade into obscurity as the major players made their presence known.

Although powered by a diminutive 246cc motor, the Lilac touted several features of a full size machine.

With an engine layout similar to a Moto-Guzzi, the Lilac even gave the illusion of a larger motorcycle.

The simplistic engine is dwarfed by the bulbous fuel tank hanging above. An enormous flywheel is housed in the cylindrical casting, helping to smooth out the operation.

The transverse V-twin nestled in the tubular frame was an exercise in simplicity, but provided more smoothness than others on the sales floors.

The heavily ribbed cylinder castings were topped by a "Bavarian-esque" valve cover.

Unlike the chain driven models offered by the competition, the Lilac was fitted with a shaft to deliver the horsepower to the rear wheel.

With a speedometer as your only gauge, instrumentation was sparse, and the warning lights offered little in the way of assistance.

A simple drum brake was fitted to each wheel, but provided ample stopping power for this lightweight machine.

The large, boxy seat was filled with enough foam to comfort the rider over the roughest of terrain.

Enormous turn signals flanked the simple headlight to give plenty of warning to fellow drivers.

Each side of the enormous fuel tank came equipped with a full size knee pad.

Continuing the "simplistic" theme, the handgrips were devoid of frills, but were molded with the manufacturers name.

1965 Kawasaki B8-T

Owner: The Jim Kersting Family Collection

Kawasaki built their first motorcycle as the Kawasaki Aircraft Company in 1949. It was not until 1960 that Kawasaki would build their first mass produced cycles. The first units were propelled by a 123cc motor that produced 8 horsepower. Later models included a 498cc vertical twin, and an 85 and 246cc version of the new two-stroke motor.

The Kawasaki Aircraft Company was formed in 1937, but would not mass produce motorcycles for many years to come.

The B8-T was one of the earliest machines to be sold by Kawasaki in this country. Perhaps crude by today's standards, the B8-T offered a variety of features that made it seem rather exotic when compared to other transportation available.

Remaining examples of early Kawasakis are rather hard to come by. Although not in perfect condition, this specimen is an accurate rendition of the original offering.

Right: Hidden beneath the sheet metal housings reside simple coil-over springs to provide the rider with a comfortable ride.

Typical for the period, the headlight nacelle housed not just the headlight, but the ignition switch, and speedometer.

The long, flat seat provided space for two, but made for a ponderous vehicle when being ridden two-up.

The stamped steel swingarm was up to the task of keeping the minimal horsepower in line.

1965 Marusho S.T. 500

Owner: The Jim Kersting Family Collection

The Marusho borrowed heavily from BMW in the design of its engine layout. The Marusho factory made its home in the dwelling previously occupied by the Lilac Motorcycle Company, and only produced these machines for a period of four years.

With its efficient design, and no-frills assembly, the S.T. 500 was actually a competent machine in its day.

Right:
A casual observer could easily mistake the Marusho for a BMW, due to the similarity of the two motors.

Below:
Along with all the other BMW styling clues, the finned valve covers played a major role.

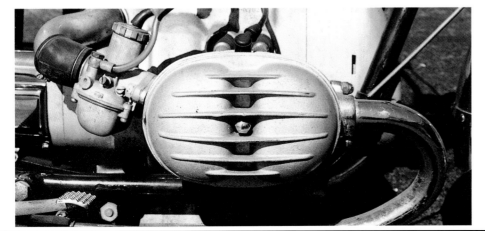

Although the casting may not be as clean as a BMW's, you cannot deny the common design of the Marusho's flat twin layout.

Not only is the layout similar to a BMW's, the exposed pushrods do nothing to dispel the cloning efforts of Marusho.

Each of the Amal VM 22 carburetors drew its breath through a curved intake manifold.

When the ride got a little wild, the rider could always use the tank mounted knee pads to help himself stay on top.

Not only was the S.T. 500 powered by an engine similar to a BMW, it was also produced with a shaft drive.

The circular tank badge stated in plain English that it was produced anywhere but Germany.

The simple speedometer was the only instrument supplied to monitor your speed. Although a top speed of 120 mph is shown, the owners manual stated that only a top speed of 100 mph is obtainable.

1966 Honda CB-77 Super Hawk

Owner: Chris & Jennifer Christensen

The stylish model on the cover of the 1966 Honda sales brochure beckoned you to "Roll out the fun." The lineup for that year included fourteen different models, and the CB-77 was only one notch below the Black Bomber, which displaced 139 cc's more than the Super Hawk. Included in the list of standard equipment were, electric start, adjustable foot pegs and rear shocks, and double leading shoe brakes. Honda was beginning to set the pace at every level of motorcycling, and the Super Hawk was one of the harbingers of things to come from the big factory across the sea.

Although displacing only 305cc's, the CB-77 delivered commendable performance, and gave the illusion of a much larger machine.

The black T-handle allowed the rider to vary the tension of the steering head to his or her liking.

In contrast to the color matched headlight assembly, the side covers were painted in a plain silver.

Encapsulated in a singular face were the speedometer, tachometer, and vertical odometer — making for a compact delivery of the pertinent data.

While still remaining rather flat, the seating of the CB-77 did feature some nicely sculpted details, which offered the rider and passenger more comfort.

The narrow handlebars were shaped to bring the grips into the rider's hands with a comfortable, upright riding position.

Adjustable rear shocks were among the standard amenities found on the Super Hawk. This feature allowed the rider to tailor the ride to meet his needs.

The combination headlight, instrument pod, and ignition switch was color keyed to the body color of the CB-77.

According to the factory sales sheets, the 305cc engine in the Super Hawk delivered a top speed of 99 miles per hour, while providing an average fuel consumption of 100 miles per gallon.

The view from the rider's perspective included the seamed fuel tank, along with simple instrumentation. The additional brightwork added to the illusion of being seated on a luxury machine.

1966 Honda Sport 160

Owner: Chris & Jennifer Christensen

For the 1966 model year, Honda offered three different variations in the 160cc class. Positioned between the 90cc models, and the more powerful 305's, the 160 was just right for the novice or experienced rider. The Sport 160 was the best all around choice of the three variants available. The Touring 160 was aimed at the boulevard gang, while the Scrambler 160 set its sights on the on/off road crew.

The lines of the Sport 160 were clean and classic. By combining all the latest features, Honda created a bike for the masses.

Both sides of the chrome tank were emblazoned with the winged logo of the Honda Motor Company.

Utilizing the motor as a stressed member left the outline of the Sport uncluttered. The twin exhaust pipes sweep smoothly to the rear, enhancing the neatness of the design.

In typical Honda fashion, the pressed steel tank is adorned with grooved knee pads for the rider's comfort.

Providing a modicum of adjustability, turning this knob allows the rider to stiffen the steering response to his liking.

The elliptical speedometer indicated the proper rpm ranges in a graphic form, rather than through the use of a true tachometer.

The overhead cam, twin cylinder motor in the Sport produced 16.5 horsepower and would carry the 160 to a top speed of 75mph.

With the use of disc brakes still some years in the future, internal shoe drums were mounted both fore and aft on the 160s.

The tubes of the forks were also kept clean with the use of steel covers in lieu of the rubber gaiters seen on other machines.

The vinyl covering of the Sports' seat would come color keyed to the paint of the machine being sold.

1966 Kawasaki W-1

Owner: The Jim Kersting Family Collection

Make no mistake, the W-1 from Kawasaki was designed with an air of British twin in its veins, but all other comparisons must stop there. Built as a touring motorcycle, the W-1 is fitted with a strong running, 624cc, vertical twin motor that will carry the rider to a top speed of over one hundred miles per hour in relative comfort. With a full tank of fuel, the W-1 weighs in at just under five hundred pounds, which helps to provide stability to an otherwise smallish machine. The build quality was considered to be one of the better examples, and sales were brisk, but not spectacular, much like the W-1 itself.

Although dressed in the trim of a British machine, the W-1 was all new technology, and pure Kawasaki inside.

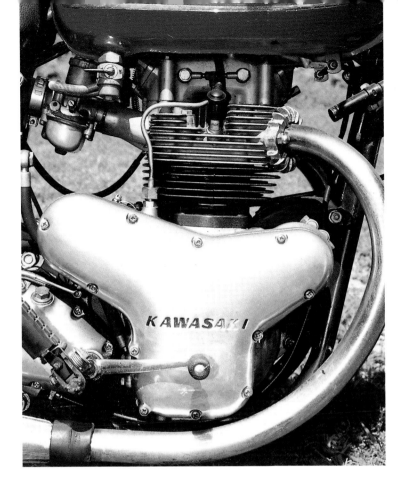

Left:
The 7.9" front drum did a respectable job of slowing the W-1, and was assisted by a seven inch brake at the rear wheel.

Far Left:
The large aluminum casting did a fine job of mimicking a British pre-unit constructed motor, the transmission is actually a separate entity.

The Nikko horn was installed at the factory and gave a solid blast, in contrast to the flat noisemakers found on lesser machines.

Even with the advent of more power and features, the pillion remained flat and featureless.

Top Left:
The British flavor starts up front with the application of the rubber fork gaiters covering the downtubes of the fork.

Top Right:
The vertical twin engine of the W-1 displaced 624cc's and provided smooth delivery of its 50 horsepower; however, it was most comfortable at highway speeds.

Bottom Left:
With more and more conveniences making their way onto motorcycles, it was no surprise to find the choke lever within easy reach of the rider's thumb.

Bottom Right:
The cast badges mounted to the side of the fuel tank featured the earlier flag of the Kawasaki Aircraft Company against a modern swirl of finely formed ribs.

1966 Suzuki T10

Owner: Martin Horn of Aero P.A. Cycle

Despite Suzuki's earlier attempts at building a motorcycle for export to the U.S., the T10 was the first to be built in sizable numbers. Predecessors to the T10 included the Colleda TT of 1956. In order to satisfy the growing needs of the American consumer, the T10 provided a much smoother powerplant that also produced less noise than the previous machines. Not only was the T10 smoother and quieter, but it sported electric starting, and a hydraulic rear brake. A new level of safety and convenience was achieved by adding these two features to an already capable form of transportation.

By joining several convenient features with the look of a larger cycle, Suzuki finally hit on a winning combination, which resulted in a substantial number of units exported to the United States.

With far more sheet metal being incorporated into the design of the T10, the stigma of the scooter image was avoided.

Pressed sheet metal technology was used extensively on the T10, including the drive chain enclosure.

The overly optimistic speedometer is housed within the headlight nacelle, and is the only instrumentation found on the T10.

The 246cc, two-stroke engine was capable of propelling the T10 to a terminal speed of 88 mph according to the factory specs.

Captured between the tubes of the hydraulic front end is the flat-bottomed headlight housing, and this design was found on several motorcycles of this era.

Top Right:
Trimmed in a typical 1960s fashion, the fuel tank was adorned with chrome sides and knee pad inlays.

Bottom Right:
Another effort to simplify the use of the T10 was found in the hinged rear fender, making the chore of rear wheel repair a much simpler affair.

Braking at both ends was provided by internally expanding, mechanical type drums.

One of the more mundane features of the T10 was the flat, two position saddle.

Offering little in the way of comfort, the handgrips were the thinnest of rubber, but followed the simplistic theme of the T10's controls.

The stylish chrome treatment went beyond the gas tank and was applied to the covers of the side panels as well.

1966 Yamaha Big Bear Scrambler

Owner: Chris & Jennifer Christensen

With the thrill of a World Grand Prix victory still charging through their veins, Yamaha introduced the YDS3-C as "The Hottest and very newest 250 in the Yamaha line." Boasting an entirely new oil injection system and twin carburetor engine, the Scrambler was sure to draw some fresh riders into the fold. Although displacing only 246cc's, the two stroke motor was promoted as having ample energy to propel a rider across hill and dale.

By utilizing a 4.1 gallon fuel tank and high mounted exhaust pipes, the Big Bear offered miles of fun on the open trails. With a MSRP of only $849, the Big Bear was seen wherever motorcyclists gathered.

Featuring a two stroke, two cylinder, two carburetor engine, the Big Bear had more horsepower while delivering tremendous fuel economy.

Above:
The dual, high mounted exhaust pipes were fitted with chrome heat shields to protect both the rider and the environment from being scorched.

The spark arrestor at the end of each exhaust tip could be easily removed for a small boost in power while riding on the street.

The circular face of the instrument panel provided the rider with all the warning lamps and engine management readings he would need.

Left:
The Big Bear featured Daytona-type, inner expanding brakes at both ends and claimed to provide superior stopping power to any other type in the world.

Below:
By combining the 4.1 gallons of fuel with the rating of 75 miles per gallon, the Big Bear owner could look forward to a long day perched atop the saddle.

Above: The front hydraulic shocks were aided by the coil springs, and promised to deliver a comfortable ride no matter how rough the roads got.

Right: Yamaha's new Autolube system eliminated the need to pre-mix your fuel and oil. The new system automatically delivered the proper amount of oil to the fuel tank, where the two fluids could mingle in an efficient fashion.

1967 Honda CB-450 Black Bomber

Owner: Chris & Jennifer Christensen

After spending their early years building a reputation as a small bike company, Honda built the CB-450. Powered by a bigger motor and assembled on a full size chassis, the Black Bomber put Honda on the map as a real motorcycle builder. The design of the Black Bomber is reminiscent of the British offerings, and did nothing but help Honda sell huge quantities of the 450 to the eager U.S. market. Within a few years of the Black Bombers debut, Honda would change the world of the motorcycle forever.

British styling clues abound on the Black Bomber, delivering an unbeatable combination of style and substance.

The view from the saddle took in the gently flowing sheet metal of the graceful, yet purposeful fuel tank.

Pumping out 43 horsepower, the vertical twin motor of the Black Bomber set a new level of performance for a Japanese motorcycle.

The adjustable steering damper was fast becoming standard equipment on machines from Honda, as well as the other Japanese manufacturers.

The coil over shocks of the CB were enclosed for neatness, and helped to keep the lines of the Bomber uncluttered.

The minimum rise of the handlebars provided a sporty, yet comfortable, riding posture for the rider.

At each end of the handlebars lay the simple molded grips. In these grips, the rider now held a high performance machine in his hands.

No longer a simple, flat place to sit, the CB-450 was given a saddle that allowed the rider to sit "in" it versus feeling perched on top.

With an indicated redline of 9500 rpm, the 450 once again raised the ante in the emerging big bike market.

Beneath the sculpted fuel tank resided a potent motor, cradled in a simple, single downtube frame.

The Black Bomber proved to be so popular that plastic kits were sold to those perhaps not old enough to ride the real thing.

The front forks were assembled with covered sliders and held a painted silver fender between the tubes.

1967 Honda CL-77 Scrambler

Owner: Chris & Jennifer Christensen

A wide variety of models continued to make up the Honda line for the 1967 model year. *Cycle World* magazine hung the "Gentleman's Scrambler" moniker on the CL-77 for its ease of operation and the comfort it offered the rider. Never really designed for serious off-road miles, the Scrambler was more at home on the street. The semi-aggressive tread pattern did offer a modicum of extra grip in the dirt, but was no match for the "knobbies" found on true dirt machines.

The upswept exhaust system offered far more ground clearance than required, and often posed a burn risk to the ill-equipped passenger.

Unfettered by the high mounted exhaust, the right side of the CL-77 seemed almost dainty when compared to other on/off road machines.

With the chrome plated heat shields in place, it was difficult to sense the evil that lurked beneath them.

The parallel twin motor breathed easily through a set of twisted outlets. Tucking the pipes in-close helped avoid burning the rider, but the passenger was often not as lucky.

Regardless of the chosen terrain, the rider was given only a speedometer to gauge his progress. Shifting was done by the seat of the pants.

To thwart a rider from altering his exhaust for added power, this ominous label was applied to the left side cover.

At the end of the heated serpent was this graceful bullet shaped exhaust collector, complete with an internal spark arrestor.

Seat design was still rather unimaginative, but a touch of white piping added an air of elegance to an otherwise slab-like saddle.

Braking was still provided by drum brakes, both front and rear.

Much like the overall design of the Scrambler, the right side of the motor is devoid of clutter.

When riding in the wild, these wire clips did their best to keep the brake from being readjusted by an errant branch of a low hanging tree.

1967 Honda Dream

Owner: Chris & Jennifer Christensen

"You meet the nicest people on a Honda" was the friendly message being conveyed to the general buying public. In an attempt to lure the non-riding public into the fold of two wheeled transportation, the Dream was built to look more like a scooter than a full grown motorcycle. Decked out in bright, fun colors, Honda hoped to get more women into the riding scene. Just mention the Dream at a party, and someone is sure to share their recollections of their first ride. Honda had truly reached a new audience with their sporty new Dream.

With the air cooled engine being the only mechanical item visible, the Dream gave the illusion of being harmless and easy for anyone to ride.

Instrumentation consisted of a single speedometer, and was probably all the typical Dream rider needed to monitor his travel.

Up front, the forks are also guarded from view with an intricate sheet metal mask.

Avoiding any resemblance to your typical motorcycle, the Dream was adorned with a variety of unusual shapes that even extended to the headlight itself.

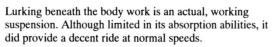

Lurking beneath the body work is an actual, working suspension. Although limited in its absorption abilities, it did provide a decent ride at normal speeds.

Beneath the cloak of the lamb beat the heart of a wolf in the 305cc twin cylinder powerplant, capable of a maximum velocity of nearly 90 miles per hour.

The hidden hardware treatment extended all the way to the rear shocks. Neatly enclosed in their steel housing, they could not frighten the innocent rider.

As if the sheet metal bodywork was not enough, the Dream was also fitted with these polished shields for an air of luxury.

To avoid soiling the trousers of a new rider, the Dream's chain was completely enclosed. This feature also removed one more mechanical monster from the neophyte's eyes.

The shape of the fuel tank implied speed and action, and was equipped with the mandatory knee pads.

Continuing with the non-motorcycle image was the brightly covered seat. The 160cc versions of the Touring model were still trimmed with an ordinary black vinyl seat cover.

Actually being a real cycle, the Dream was equipped with a steering dampener for the hard core rider.

1967 Honda Super 90

Owner: Chris & Jennifer Christensen

Amongst those wanting to ride, there were some afraid of a full sized machine, while others would rather not be seen on a mere scooter. The Super 90 was conceived in an effort to bridge the gap, and appeal to both parties. Although driven by a tiny eight horsepower motor, the Super 90 avoided the step-through look of the similarly powered scooters on the market. Weighing in at a whopping 176 pounds made the 90 easy for anyone to manipulate.

The diminutive Super 90 attempted to fill the gap between those who desired a real motorcycle and those who still wanted to start small.

With a terminal velocity of almost 65 miles per hour, the speedometer of the 90 was mildly optimistic.

The single cylinder motor in the Super 90 produced 8 horsepower at 9,500 rpm and did an adequate job of moving the rider along.

The compound curves of the handlebars provided a comfortable reach along with plenty of stability.

The plastic duckbill on the rear fender helped to keep the rider's back dry during an unplanned ride over wet roads.

There were even fold up footpegs for a passenger, but the weight of another passenger taxed the Super 90's ability to perform.

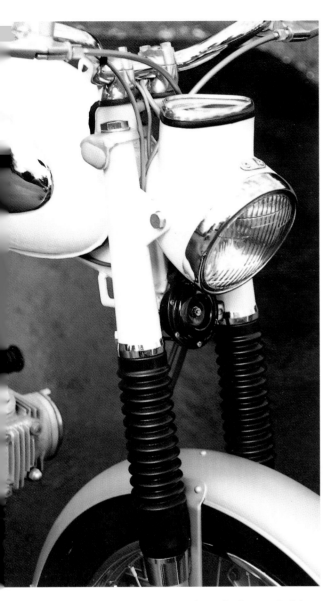

Replete with rubber fork gaiters, the front end of the 90 took on the look of a full sized British machine.

It did not take a lot of brake to slow the Super 90 down, so the tiny drum brakes seemed right at home at both ends of the chassis.

A set of real shock absorbers added to the look and feel of a bigger machine, and provided ample comfort.

In deference to the overall dimensions of the Super, the saddle seemed like a throne fit for a King.

1967 Suzuki X-6 Hustler

Owner: Chris & Jennifer Christensen

After setting a new speed record at the Bonneville Salt Flats, the X-6 rolled into the showroom floors carrying the flag of glory. A popular ad of 1967 was titled "Switch Craft", and spoke of how everyone was changing horses to ride the new Hustler. The motor was a 250cc two stroke, providing better than average acceleration. Ridden by a professional rider, the X-6 turned a 14.82 quarter mile. Not too shabby for such a tiny machine. The Hustler was one of the first production machines to come equipped with a six speed transmission.

The Hustler offered superior performance in a compact package. The two stroke motor proved to be an able partner in the construction of the X-6.

The painted, steel fenders were both trimmed by a vibrant red stripe, which lent a racy look to the high performance Hustler.

The instrument face featured both speedo and tach. The red triangle over the "5" was used as a break-in indicator, and could be peeled off once break-in was achieved.

Front suspension consisted of hydraulic struts assisted by exposed coil springs.

The tubular handlebars provided ample lift and gave the rider a comfortable reach to the molded grips.

The knee pad inserts continued to be used on the X-6 models, but their useful existence was beginning to fade.

The new "Posi-Force" lubrication system eliminated pre-mixing of oil and fuel, and claimed to provide a smoke free machine, which was quite a feat for a 2 stroke motor.

The ever popular slab seat was still in position on the X-6, but at least provided some cushioning for the rider.

1969 Honda CB750 Sandcast

Owner: Ray Landy

Once Honda had convinced the world that you meet some nice people on their two wheeled machines, they decided to up the ante. While motorcycles had been growing slowly into real machines, they seldom were powered by more than two cylinders. With the introduction of the all new, in-line four CB750, Honda changed the face of motorcycling forever. The first of the new breed were delivered with sand cast engine cases, making them the rarest of the breed. The single overhead cam engine reset the scales in the performance world, and sent the other manufacturers scrambling to catch up. Honda would repeat its earth shattering creativity again and again over the years, but the CB750 was still the first "superbike" to hit the streets of America.

The cycle that changed the world. The first CB750's had motor castings that were done in sand, but were later updated to more conventional methods.

Four cylinders, all in a row — Honda had achieved all new records for horsepower and sales with the introduction of the CB750.

The sidecover stated the size of the new CB, and told the rest of the industry to get busy.

Still a classic today, the CB750 set the performance bar up several notches and was untouched for years as the superbike King.

A single disc brake was mounted to the spoke wheel up front, and was considered quite radical at the time.

The single overhead cam motor boasted unheard of performance specs, and lived up to the claims.

In addition to the painted headlight housing, the front forks were fitted with gaiters to keep the sliders clean.

With the warning lights integrated into the faces of the instruments, the CB was also an exercise in simplicity.

The four-into-four exhaust split into pairs, and the top tube was fitted with a heat shield to protect the passenger's legs from burns.

1971 Honda CL-350

Owner: Steve Henkel

The early part of the 1970s saw the Scrambler designation still in force. Equipped with both street and off road hardware, these machines were supposed to provide the ultimate in versatility. While handling the rough terrain better than a Norton Commando, the Scramblers still offered little in the trails department to handle the really tough hills and dales. They were actually far better suited for use on the streets, and that is where most made their home.

With its high mounted exhaust system and energetic powerplant, the CL was ready to take on all comers.

The 325cc motor cranked out 33 horsepower and gave the CL a top speed of 100 miles per hour.

The tubular steel swingarm's motion was damped by the application of mildly adjustable shocks.

The slotted heat shields covered the tucked-in exhaust, but offered little protection from the searing fumes that left the engine.

The sturdy handlebars carried an additional cross brace for the rugged adventures of the Scrambler rider

Drum brakes were still in vogue and provided sufficient slowing power for these lightweight machines.

The overall shape of the fuel tank seemed to be much larger than the 2.4 gallons it held within.

Giving the appearance of a double barreled shotgun, the exhaust outlets had a bark worse than their bite.

The cockpit remained simple, and carried only the bare essentials.

The final segment of the exhaust system was shielded with another pair of nearly useless perforated heat shields.

1971 Yamaha R5 350

Owner: Rick Youngblood

By improving on the highlights of the past models, and incorporating some new features, Yamaha created a winning combination in the R5. Not only was the R5 a great machine, it sold for only $739 brand new. Weighing in at just over 300 pounds, the R5 delivered a new level of performance in an easy to handle format. Slimmer, lower, and lighter than previous editions, the R5 was winning new riders, as well as bringing current Yamaha owners into the showrooms. The R5 would continue to be improved, and Yamaha offered a two stroke machine for many years to come.

Sporting leaner lines and more horsepower, the R5 was now one of the top rated machines by the cycling press.

The two-tone fuel tank was topped off with a screw in gas cap emblazoned with the Yamaha tuning fork logo.

The deeply finned cylinder heads allowed for the power-mad twin to remain cool during performance runs.

The tubular duplex cradle chassis provides a safe haven for the narrower twin of the R5. Riders found the R5 to be a willing machine that usually started on the first kick.

The newly designed motor was dressed up with fancy castings that carried the name of the manufacturer, as well as decorative accent striping.

The early 1970s models began to show signs of more flash than previous offerings. The chrome accent strip on the side covers was one indication.

Front stopping power was supplied by a 7.2 inch drum brake.

The rear suspension layout was typical, including coil over shocks and another 7.2 inch drum brake to aid in slowing the R5.

Below Left: Instrumentation was simple, including an exuberant speedometer. The top speed of the R5 peaked at less than 100 miles per hour, so the 120 indicated was merely a dream for the R5.

Below Right: Each of the alternately firing cylinders takes its breath through an independent carburetor fixed with faux velocity stacks.

1973 Honda CB350

Owner: Rick Youngblood

Amongst the proliferation of models offered by Honda, the CB350 twin was not as well received as others. When asked, most enthusiasts remember the in-line-four 350 that led to the 400/4, but the vertical twin draws a blank. This lack of recognition did not make the two cylinder version a lesser machine. As a beginner's cycle, the CB350 excelled. It offered a simple, light weight package that was hard to resist.

With its full sized features and light weight, the CB350 twin was a great choice for those waiting to join the two wheeled race.

Above Left:
In the old days, a motorcycle with a flat battery was not dead in the water. You could always put the kick starter into action.

Above Center:
The sidecovers were simply trimmed with vertical slots for a touch of flair.

Above Right:
The front end was also quite normal with its drum brake and hydraulic forks.

The hinged, locking fuel cap became an important feature during the fuel "crisis" of the 1970s.

The enormous tail light gave more than adequate warning of your intentions.

Above Left:
Capable of traversing long miles on the highway, the 350 was fitted with a wide seat to coddle the rider and passenger.

Above Right:
The sweeping tank graphics suggest a long curving highway, and would become a classic trademark for Honda's of this era.

Bottom Left:
The vertical twin motor was compactly designed to keep the machine narrow and easy to maneuver.

Bottom Right:
As was still the typical practice, the rear was slowed by a drum brake and suspended by coil over shocks.

1973 Kawasaki H-2

Owner: Jim Kersting Family Collection

Continuing their tradition of building ultra fast two stroke machines, Kawasaki rolled out the H-2 alongside its 500cc sibling, the H-1. An additional 250cc's made the H-2 an extremely quick motorcycle. Equalling the lightning fast response from the throttle was the cat-like handling from the three cylinder models, or triples, as they would soon become known. Their only drawback was the narrow powerband (the rpm range where the power comes on and drops off), which was a lot like a light switch. Much like a light switch, you were either cruising along in the gutless lower rpm's, or catapulted into warp speed by the free spirited higher revs. This was a difficult ride for inexperienced riders.

With a wide selection of curves and shapes, the H-2 had a slightly ungainly appearance.

The redline may be "only" 7,500 rpm, but the H-2 packed most of its speed into the upper third of that range.

Despite its awkward styling, the H-2 performed like a sweet beast when the throttle was twisted open.

Paint selections and graphics were rather typical for a cycle out of the 1970s. You could find a wide variety of candies and metalflakes when shopping the showroom floors.

The wide, flat seat was thinly padded and did not offer a lot of comfort to the all-day rider.

Although similar to the tail that would grace the Z-1, the H-2's was slightly square and offered less visual appeal.

Top Left:
With modern technology finally finding its way into cycles, disc brakes were being used in more-and-more cases.

Top Right:
The three cylinder motor sat nestled in the safe harbor of a double cradle frame and was deeply finned to help keep it cool.

Bottom Left:
The motor was actually quite compact when compared to the punch it could deliver.

Bottom Right:
The business end of the H-2, the side most often seen by the uninformed rider of lesser machinery.

1973 Kawasaki Z-1

Owner: Al Steier

The world of motorcycling had hardly recovered from Honda's CB750 bombshell when Kawasaki decided to raise the stakes. At the unveiling of the 903cc Z-1, engineers from the other manufacturers could almost be heard clamoring for their calculators. Many of the earlier Kawasakis had been built around torque producing two stroke motors. As great as they were for scorching away from stop lights, their scope was limited to that field of endeavor. The four cylinder powerhouse squeezed between the rails of the Z-1 made power at nearly every point on the tach, and reached a peak of 82 horsepower at 8,500 rpm. A new standard had been set, and Kawasaki was proud of it.

Although displacing nearly 550 pounds when tanked up, the Z-1's lines gave it a look of speed, even while at rest.

Each cylinder of the in-line four exhaled through its own tube. This improved breathing played a big part in the performance of the Z.

Above Right:
Nestled between the front forks is a headlight measuring seven inches in diameter. All the better to see the speed limit signs at night.

Bottom Right:
The badge on the side cover said it all. "Double Overhead Camshaft" told the rest of the gang that you meant business.

Above:
The four lungs are fed through Mikuni VM28 carburetors and do a fine job once the Z-1 has warmed up to operating temperature.

The rear shocks are fitted with large reflectors to give the slow moving competition a brief glimpse of your passing.

In addition to the eight inch drum brake on the rear wheel, the Z-1 was hauled down to legal speeds with a single hydraulic disc up front.

Below:
In between the large, easy to read speedometer and tach, you find a bank of idiot lights to keep you informed of serious trouble.

At the ends of each exhaust pipe, you find the upswept megaphones that do their best to quiet the monsterous noises that escape from their orifices.

The wide, flat seat offered lots of comfort for those long highway jaunts, but nothing to keep the rider in place during rocket-like acceleration.

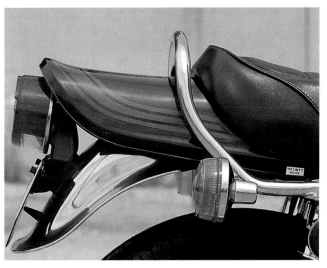

The Z-1 introduced the world to the duck tail fender. This design would become a trademark for all Z-1's to follow.

The all steel fuel tank held 4.75 gallons of go juice and its wrap around graphics added to the graceful lines of the metal beneath.

1974 Honda CB200

Owner: Dan Davis

Ever since Honda's earthshaking introduction of the CB750 in 1969, there was less attention paid to the introductory rider's needs. More time and effort was being spent on bigger and faster bikes, while the smaller machines began to languish on the showroom floor. In deference to the market, the CB200 offered simple, economical transportation for people not yet prepared to step up to the big leagues. Small and agile, the CB200 was the perfect bike for beginning riders. The Dream had been Honda's earlier foray into the entry level market, and had won the hearts of many riders since then. The CB200 provided the next logical step for someone graduating from their Dream into the world of real motorcycles.

The diminutive CB200 offered a lot for a low entry cost and was employed by many as a stepping stone to bigger rides.

The twin cylinder motor sat neatly within the single down tube frame, and access to the valve adjustment covers was superb.

The angular lines of the tank were accented by the plastic divider that was applied to the top surface of the sheet metal.

The extremely narrow tank and handlebars let small riders feel at home, while providing plenty of control.

Above:
Out back, a drum brake was on duty to assist the disc up front. The shocks were simple coil over units, but they did an admirable job of soaking up small irregularities.

Right:
The sidecovers mirrored the shape of the fuel tank and were trimmed with a simple alphanumeric badge.

Above:
In lieu of the standard drum brake, the CB was actually stopped by a single disc brake on the front wheel.

Although a simple cycle, the CB200 was equipped with separate instruments for both speed and rpm.

1974 Kawasaki H-1

Owner: Jim Kersting Family Collection

Perhaps overshadowed by the bigger H-2 available at the dealer, the H-1 still had a loyal following of eager two stroke devotees. Displacing 250 fewer cc's than it's sibling, the H-1 did not produce the same level of thrills, but was a competent machine nonetheless. With a graphics package reminiscent of the Z-1 superbike, the H-1 hoped to make a mental connection with the in-line four model. It would only be one more year before the government forced two stroke machines off the market, so the H-1 enjoyed an enthusiastic, albeit brief, period in the limelight.

The profile of the H-1 was much sleeker than the H-2 and retained a stronger resemblance to the Z-1.

The industry standard drum brake was in place on the aft portion of the H-1.

The three cylinder motor boasted a displacement of 500cc's and accelerated like a jet plane when the rider hit the powerband.

The installation of disc brakes was becoming common, but only on the front wheels of most motorcycles.

The rear view included three shining exhaust ports, a narrow tire, and an unusually shaped tail light.

In stark contrast to the H-1's performance, the seating remained rather dull and did little more than give the rider a place to sit.

Three basic warning lights were flanked by two simple instruments; however, these provided the pilot with enough feedback to keep him on track.

1974 Yamaha RD 350

Owner: Steve Searles

With its family tree reaching all the way back to the race-track-only TR3, the RD can be proud of its lineage. Right after the turn of the decade, Yamaha introduced a street going version of the TR3 in the R-5. Although somewhat tamer than its racing brother, the R-5 proved that fast things come in small packages. Years of improvements led to the RD 350. With a curb weight of 310 pounds, and nearly 30 horsepower on tap, the RD makes for a potent package. It was a capable machine that could be pressed into road racing duty, or could gobble up hundreds of miles on the interstate. It was truly a multi-faceted vehicle amongst a field full of single purpose hardware.

The RD 350 was touted as the best handling street bike sold in America by much of the motorcycle press.

A truly fine motorcycle must do more than go fast and handle well. The disc brake at the front wheel assured quick, yet safe stops aboard the RD.

The "Torque Induction" system was just one facet of the RD's easy starting, quick running nature.

The black finish added an air of elegance to the tiny giant-killer of a motor.

The balance of the RD's braking was done by a drum on the back wheel. The adjustable shocks gave a predictable ride.

Much like the overall design of the RD, the instrumentation was basic, yet informative.

Tipping the scales at a svelte 310 pounds, the RD delivered spectacular performance from its miniature stature.

The steel tank held 3.2 gallons of fuel and would carry the rider nearly 150 miles under peaceful riding conditions.

Although displacing only 347cc's, the RD did a fine job of out performing many of the larger machines roaming the streets in 1974.

A sea of red plastic lenses greet the rider of anything less than an RD 350.

1975 Honda CB400/4

Owner: Rick Youngblood

At its debut in 1975, the CB400/4 was really nothing more than an upgraded 350/4 from the year before. What the 400 lacked in cutting edge technology, it made up in fun. The 400 was stirred with a six speed transmission, and was one of the smoothest running bikes on the market. It would be handily trounced by the two strokes from Kawasaki, but it brought a new level of sophistication into the fold of small, four cylinder cycles. With a more aggressively styled tank, and serpentine exhaust, the CB400 provided lots of excitement before it was even started. With a red line of 10,000 rpm, there were still plenty of thrills to be had.

The visual appeal of the CB400 was tough to deny. With the combination of a new, angular fuel tank and a flowing exhaust system, there was almost too much for the average human to absorb.

With the front caliper gripping a single disc, braking was as confident as the performance.

The "Super Sport" moniker was confirmed instantly by the writhing exhaust pipes and their single megaphone exit.

Final drive was still chain driven and few manufacturers would deviate from that form for several years to come.

Far Right:
On the rare occasions the rider would actually slow down, his intentions were made public with the large stop light.

Right:
The sporting intentions of the 400 were proven by the mild rearset footpegs, and the tidy linkage connecting the lever to the transmission.

Since the CB400 was fitted with a four-into-one exhaust, the left side was devoid of the clutter of a megaphone.

The instrument cluster fell between the low rise of the tubular handlebars and made for an appealing command center.

Rear wheel braking was provided by a drum that did its best to bring the 18 inch wheel to rest.

The aluminum oil filter housing was extremely easy to reach, but the hex head retaining bolt was prone to rounding off the first time you attempted to remove it.

1976 Honda GL1000 Limited Edition

Owner: Ray Landy

The motorcycle world had barely recovered from Honda's first bombshell in the CB750, when they rolled out their latest machine. The GL 1000 first appeared in 1975, and quickly set the standard for touring bikes. Its opposed four cylinder engine was like nothing seen before in a motorcycle frame, and it provided a smoothness never known to riders before. The "fuel tank" was actually a hiding place for electronics and storage, while the real tank was lower in the frame to aid in balance and handling. For 1976, Honda rolled out the Limited Edition model of the GL. replete with gold wheels, and special badges, the Limited was only reproduced 2000 times, in contrast to the hundreds of thousands of "Wings" that would eventually take to the highways of the world.

This limited edition Gold Wing appeared in the second year of production and was trimmed with a variety of special features.

The limited edition of the GL1000 came complete with a leather key fob that concealed the key inside when not in use.

The enormous radiator was present to keep things cool. Honda did their best to tuck it out of sight.

The rear wheel was fitted with a full sized disc brake and was placed behind the dual shotgun exhaust tips.

A fuel gauge was supplied; it lived in the top of the "fuel tank."

Instrumentation was simple, yet complete. Even a temperature gauge was included in the set.

Just as in the first year, the tank is actually a storage area, and a place to conceal electronic components from the elements.

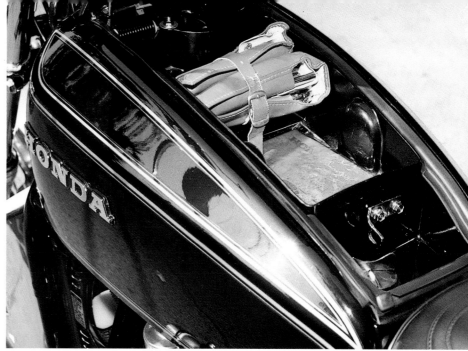

The storage area in the limited edition was the same, but came complete with a leather pouch for both the tools, and the owner's manual.

The opposed four motor from the first Wing was still in place and continued to be one of the smoothest powerplants on earth.

To keep the final drive as silent as the motor, shaft drive was the obvious choice.

1978 Kawasaki Z1-R

Owner: Jim Goebel

Just because the release of the Z-1 set the motorcycle world on its ear was no reason for Kawasaki to rest on their laurels. In lieu of a completely new model, they decided to create a special edition of the original Z-1. Dubbed the Z1-R, the original machine was outfitted with numerous improvements. Lurking beneath the all new, silver-blue bodywork was a new 1015cc version of the original 903cc motor. This new variant expelled its gases through a factory installed four-into-one header. The use of fiberglass components helped to lower the weight of the Z1-R, and it was now capable of turning a quarter mile in less than 12 seconds, making it very special at the track.

Dressed in all new bodywork and exclusive paint, the Z1-R cast a large shadow on the existing ranks of the superbike class.

The four-into-one exhaust was bolted on right at the factory and made a big difference in the performance arena.

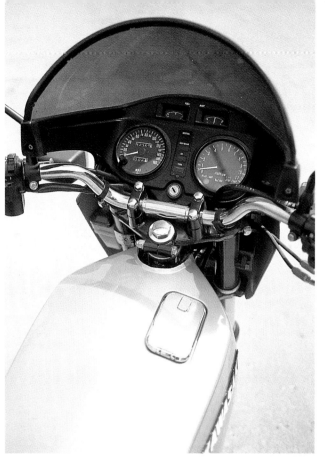

Resembling the cockpit of a small airplane, the Z1-R was equipped with the typical gauges plus a fuel gauge and amp meter to monitor conditions.

Beneath the angular steel tank lived a 1015cc monster of a motor that would take its toll on the competition.

The heavy use of lightweight fiberglass reduced the overall heft of the Z1-R, allowing it to reach new lows at the drag strip.

Each of the mated sections looked like a single thought as one line led into the next to create a flowing, yet focused look.

Top Left:
Sharp creases in the fairing averted the airflow around the rider's shoulders ... and looked great.

Top Right:
Each of the Z1-R's three disc brakes were drilled for lightness and additional stopping power in the wet.

Bottom Left:
While much of the Z1-R was all new, the rear suspension remained hydraulic with coil spring assistance.

Bottom Right:
The crisp styling lines of the Z1-R were evident anywhere there was body work.

1978 Kawasaki Z1-R TC

Owner: Tom Turner/Woodfield Cycle

With the introductions of the Z-1, and Z1-R, Kawasaki was on a serious performance bent. Not ready to stop and smell the roses, they went for all the marbles with the Z1-R TC. By taking the already potent performance of the Z1-R, and bolting on a turbocharger, Kawasaki sent the horsepower rating into the red. The bodywork was easily recognized as that of the Z1-R, but was now clad in a sinister black with orange and yellow graphics. As soon as the new TC hit the streets, it was awarded with the horsepower King's crown. Not wanting to be left behind, the other manufacturers raced to their drawing boards to create their own puffer bikes.

The TC was easily distinguished from its normally aspirated brethren by either the garish paint or the enormous exhaust system.

Bolted to the cast, seven spoke wheel were dual disc rotors to haul the TC down from insane top speeds.

A virtual hornet's nest of tubing — the exhaust gasses were collected into a single manifold before being allowed to blow past the turbo impellers.

The 1015cc's of blown motor provided lots of power with minimal noise. The turbo unit soaked up most of the racket before it could sneak out of the exhaust.

Bottom Left: The standard Z1-R gauges were joined by a huge boost indicator that announced the pressure your motor was undergoing at any given moment.

Bottom Right: The single carburetor drew every breath through this enormous air cleaner before passing it into the intake manifold.

Even if you didn't know a turbo from a train, the "TC" on the sidepanel did little to disguise the intent of this two wheeled projectile.

The cast aluminum heat shield did what it could to dissipate heat before scorching the rider's legs.

The same angular tank from the Z1-R was in place on the TC, but was also dipped in black and striped in orange and yellow.

The final exit of the spent exhaust gasses passed through this humongous bologna sliced tip.

1978 Yamaha XS-750

Owner: Rick Youngblood

Looking back through the pages of motorcycling's history, you find only a handful of three cylinder motors being used. The Triumph Trident certainly comes to mind, as do the "torquey" two strokes from Kawasaki. Almost a forgotten hero, the Yamaha XS750 filled an important slot in the annals of time. Featuring an in-line, three cylinder motor, the XS was quick, quiet, and extremely smooth—a hat trick of qualities not found on many machines of the era.

Featuring a host of convenient options, the XS was well equipped as a cruiser, or long range touring machine.

Besides being narrow, the triple cylinder motor provided an extremely wide powerband without a fuss.

The extremely narrow triple allowed the XS to be ridden by riders with less than gargantuan dimensions.

The view from the pillion included the classic pinstriping on the tank and an easy-to-read instrument cluster.

Bottom Left: To match the no nonsense layout of the XS, the side covers went unmarred by frills, but did their job to cover vital components.

Bottom Right: At long last, manufacturers were installing disc brakes on both wheels. This came in handy since motorcycles were reaching new levels of straight line performance every year.

The small tail section was not in place for style only. It held small items within after lifting the seat.

In addition to the single disc brake on the rear wheel, the front rim was equipped with dual disc brakes to ensure smooth stopping in any condition.

One of the major distinctions of the XS was its shaft drive. This choice made a big difference in overall smoothness and ease of maintenance.

Right:
Falling into the category of "standard" motorcycles, the XS750 did an above average job of doing many things.

1981 Honda CM400A

Owner: Matt and Bonnie Jonas

No matter how neat and easy motorcycle riding had become, there were still some that felt bothered by the need to shift the transmission manually. Moto Guzzi had built the 1000cc Convert in the middle 1970s, but it was a behemoth that scared away the heartiest of beginning riders. Honda addressed the need for shiftless riding, at a novice's level by rolling out the CM400A. Having two forward speeds for changing road conditions, the CM was easily shifted with a click of the foot lever. There was no clutch to engage, so the fear of riding was forever eliminated for those timid few. As it turns out, there was not enough demand for the automatic model, and it was soon phased out of production.

Done up in what has become known as the "cruiser" mode, the CM400A kept the rider close to the ground and provided a decent place to spend the day.

Left: Much like its four wheeled counterparts, the CM had a parking brake in lieu of leaving the gearbox engaged.

The side covers made it obvious to everyone that you were riding a cycle that needed no shifting.

The teardrop tank epitomized the "cruiser" genre, but limited your fuel capacity.

The bank of indicator lights between the instruments was now standard equipment on nearly every Honda, but the Hondamatic gear indicator was a specialty item.

In an effort to keep the rider close to the ground, this stepped seat was employed. The large, well padded saddle was rather comfortable regardless of its height.

Another one of Honda's trademarks was the Comstar wheel. Composed of several sections of lightweight steel, they offered tremendous strength without the penalty of additional weight.

Below:
A confidence inspiring disc brake was on duty up front, allowing anyone to slow in an orderly fashion.

Above Left:
The vertical twin motor made adequate power for a beginner's bike, but would quickly be outgrown by most riders.

Suspended beneath the engine was an additional heat sink to disperse the additional warmth generated by the automatic transmission.

1981 Yamaha 550 Seca

Owner: Barbara Pugh

The early 1980s saw a huge proliferation of new models, market segments, and riding styles. The 550 class was chock full of different styles, and you could easily pick a machine to fill your needs. Yamaha's answer to this portion of the field was the 550 Seca. With an in-line four mounted in its frame, it did not offer anything radically different. What it did offer was "swoopy" bodywork, and splashy graphics. Scorned as a woman's bike by many male riders, the Seca never sold in vast numbers. If these Seca bashers had ever thrown a leg over the seat, they would have found a competent machine that handled well, and delivered power near the head of its class. Yamaha saw little need to compete at this level, and soon bumped the Seca to the 750 rating.

Bedecked in its flowing sheet metal and unmistakable graphics, the 550 Seca truly stood out amongst the other machines in the class.

A single disc was applied to the front wheel, providing most of the stopping power.

The fairing did more than protect the rider from the wind, it gave the designers a chance to flex their muscles by creating a truly unique enclosure.

Above Center:
The in-line four of the Seca was compact and provided a respectable quantity of power.

Cast aluminum wheels were all the rage, but Yamaha adopted a drum brake on the rear wheel to save weight and money.

All major functions were monitored by this comprehensive set of gauges.

The lines of the fuel tank made it seem as liquid as the fluid held within.

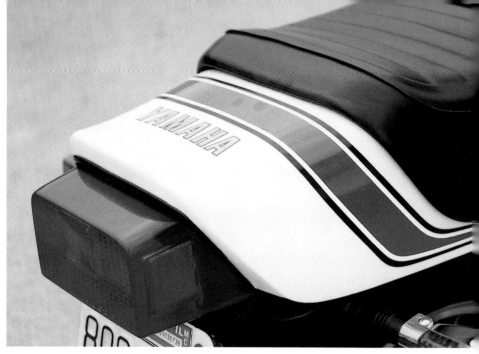
Behind the seat resided a small tail section which allowed for the storage of small items.

The mild step in the saddle helped to keep the rider in place and the strap gave a passenger some hope of staying aboard.

While many of the sportbikes were shying away from the use of chrome, the Seca bucked the trend and hung a brightly chromed fender over the front wheel.

1982 Honda CBX

Owner: Steve Searles

The six cylinder CBX was launched in 1978, and immediately upset the apple cart of the sportbike universe. With 108 horsepower in the palm of your hand, and a motor wider than the Grand Canyon, there seemed to be no way it could be usurped of its title as King of the superbikes. Within a year, the other manufacturers released their own two wheeled bullets, and although not as radical as the CBX, they were lighter, faster, and possessed greater handling. Instead of killing the CBX off entirely, Honda decided to push it into the touring category. Equipped with a full faring and saddlebags, the CBX made a pleasing long distance machine. Competition soon arose in this arena as well, and the CBX found itself without a kingdom to rule. The 1982 model year would be the last for the big six.

The final iteration of the CBX did its best to be a big touring mount but proved too ponderous at slow speeds to live up to the needs of the rider.

A wonder of modern science, the CBX's in-line six was amazingly smooth and did double duty as a stressed member of the chassis.

The enormous tail light was surrounded by acres of fiberglass making up the saddlebags.

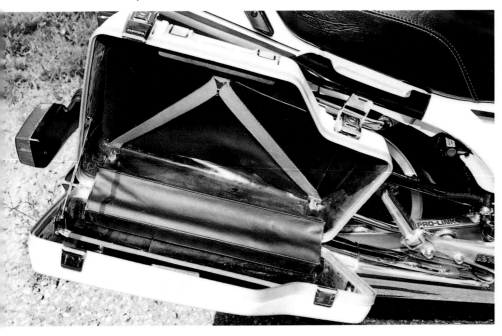

The saddlebags proved to be another deficiency of the CBX's touring garb. They were chastised for being not quite large enough to satisfy the true long distance rider, although their shape swallowed a large volume of goods.

In keeping with the nature of the CBX, the seating was made for two, and provided enough room to allow some shifting around while gobbling up the miles.

The "Pro-Link" suspension was Honda's marketing term for single shock.

The rear wheel was equipped with a standard disc brake to provide an extra dose of stopping power.

Two enormous, ventilated rotors were mounted to the front wheel, the minimum required to stop this six cylinder leviathan.

Even the sidecovers of the CBX were large, providing a great place for graphics or billboard advertising.

As indicated by the street sign, the road came to an end for the CBX in 1982. It was a techno-marvel that never quite fit in.

1982 Kawasaki KZ1000R

Owner: Al Pinkus

To commemorate their victories in the Superbike series, Kawasaki offered the masses a replica of Eddie Lawson's winning machine. Clad in official Team Green, it was tough to mistake the "R" model for anything else. Along with the vivid coloration, the Eddie Lawson Replica, or ELR, was also fitted with several performance enhancements. The addition of a small oil cooler, piggyback shocks, and a Kerker 4-into-1 header brought the ELR close to being the real McCoy. The engine remained a stock KZ1000 plant—no slouch in its own right—and would deliver a top speed of 133 miles per hour. Sold in limited quantities, the ELR remains a classic throwback to the days of the superbike.

That which was not green was trimmed in black and made for a stunning visual concept.

Looking every inch a race bike, the ELR offered great performance in wolves' clothing.

The piggyback reservoir shocks, Kerker header, and drilled disc brake make quite a trio and do nothing to tame the appearance of the ELR.

Although looking the part of racing equipment, the reservoir shocks did nothing to enhance the ride or improve handling on the street.

The leading edge of the ELR was equipped with a bikini fairing and dual disc brakes — one to break the wind and the other to brake the bike.

The gold accented wheel was paired with ventilated rotors to provide maximum stopping power.

Top Right:
The top of the fuel tank was given a decal stating the World Champion nature of the ELR and was complimented by a complete set of gauges for the rider's education.

Bottom Right:
If the screaming green paint did not get your attention, the gold shocks and racing exhaust were sure to make you stand up and take notice.

1983 Honda CX650T

Owner: Matt Jonas

The early 1980s continued to be an extremely active period for Japanese cycle manufacturers, and new models were being introduced at a furious pace. Honda's first puffer bike was the CX500T, the world's first production motorcycle to be turbocharged. The Kawasaki Z1-R TC of 1978 vintage had a Rayjay turbo bolted to a Kawasaki Z1-R and was basically hand assembled. The 1983 model year saw the CX500 grow by 150 cc's, so naturally the turbo version had more cc's as well. Along with the bigger motor, the 650 received a new graphics package that replaced the safety-glow orange and gray trim with a more refined blue and red. The turbo bikes continued to be technical marvels, but the insurance industry promptly brought an end to the era of these complex, expensive to repair machines.

Featuring a more muscular 650cc motor for 1983, the turbo gave an even more impressive performance on the streets.

The fuel cap was highly polished, and featured an enameled Honda badge.

The drilled alloy wheels were exclusive to the turbo bikes and were fitted with dual discs up front.

A third disc brake was mounted to the rear wheel and was quite a necessity when slowing the turbo from a high speed run.

The shaft drive was carried over from the non-turbo CX. With no chain to clatter and break, the turbo was nearly faultless.

Making no bones about the source of power in the frame, the CX650T proudly displayed the turbo banner on both of the exhaust outlets.

Left: The tail piece also featured an aluminum grab rail, giving the passenger a place to hold on to during the rush of turbocharged boost.

Right:
Just above the sleek turn signals sat the stalk type mirrors. The smooth operation of the CX-T provided crystal clear vision to the rear.

Below:
A far more subtle "Turbo" badge was applied to the front of the CX-T, contrasting nicely against the pearl white paint covering the panels.

Left:
The transverse, V twin motor was not the strongest kid on the block until Honda strapped on the turbo. Making vast horsepower came at the expense of simplicity, the turbo models were nightmares to service.

Below:
Back for a second year, Honda's turbo bike looked even better with the new red, white, and blue livery.

105

1983 Kawasaki GPz550

Owner: Armand Ciabattari

First making the scene in the 1982 model year, The GPz550 set new standards for middleweights. Based on the previous KZ550, the new version was treated to several basic performance tricks to achieve its muscular abilities. By boosting compression, adding a hotter cam, and increasing the valve seat area, Kawasaki created a monster. A "mini" bike no more, the 550 gave many of the 750s a run for their money. Although slightly heavier than its predecessor, the GPz more than made up for it in the speed and handling departments. About the only shortcoming the 550 had was its inability to provide a comfortable ride over long distances. Only then did the firecracker red 550 fall short of being the perfect machine for all-around application.

The year 1983 was the second year for the GPz550. It was outfitted with more flowing body panels and additional graphics.

Exceptional braking was available on the 550, due in part to the dual 226mm discs on the front wheel.

Top Center:
The mounted view takes in the sleek fuel tank and a well equipped instrument panel that included liquid crystal readouts in between the big, easy to read tach and speedo.

Above:
Just in case the front anchors were not enough, the rear wheel was mated to another 226mm disc for an extra margin of safety.

Left:
The four gallon fuel tank swept back into the sidecovers and looked more compact than its capacity would suggest.

The nicely finished bikini fairing kept a few bugs off the rider's jacket and added to the sporting nature of the little GPz.

Top Right:
The "Uni-Trak" moniker was applied to Kawasaki's mono-shock system that offered both comfort and handling to the 550's rider.

Right:
The GPz packed a lot of performance into a tidy package. The 553cc motor slung the GPz through the quarter mile in just over 12 seconds and offered quick handling to boot.

1983 Suzuki Katana 1100

Owner: Michael Mini

Introduced in 1982 as a 1000cc sportbike, the Katana found itself surrounded by 1100 machines. Although quicker than many of the 1100s in 1982, the Katana was bumped to 1074cc's for the 1983 selling season. The initial Kat was sold in silver only, but was revised to a two-tone scheme in 1983. It was also offered in a 750 cc version in 1983, and wore a suit of silver and burgundy. The radical styling of the Katana left no head unturned. Suzuki commissioned a freelance artist to do the initial design, then refined the drawings into steel and fiberglass. From the hunchback fuel tank to the two-tone seating, the Katana left little to the imagination. Sold in minuscule numbers, the Katana would gain almost cult status in later years. Katanas are seldom seen on the streets anymore, and hardly ever offered for sale.

Having the appearance of fine art on two wheels, the Katana was either loved or hated, but never failed to get people's attention.

The Katana was easily hauled down from speed by applying any combination of the 3 disc brakes fitted to the turbo wheels.

While being long in style, the Katana came up short in the comfort department. Even the adjustable rear shocks did little to improve the buck board ride of the Katana.

Anti-dive sensors were fitted to the front disc brakes to add another level of safety to the Katana.

The oddly shaped saddle did not offer a lot in the way of comfort, but who said art had to be comfortable?

The side pod carried a rotary choke lever. Two rectangular accessory switches could be installed in place of the blanks.

The wedge shaped fairing was topped off with a small windscreen that did not offer a lot of protection, but fit right in with the Katana's styling.

Top Center:
Behind the tiny windshield lies an equally unusual instrument pod. The large red indicator light told the rider his sidestand was down.

The profile of the fairing shows how little plexiglass stood between the rider and the rushing wind.

Left:
With the engine boosted in size, and now trimmed in black, the bodywork became even more obvious in contrast to the onyx motor.

1983 Yamaha 900 Seca

Owner: Matt Jonas

Seeing a need for a big, fast sportbike in their lineup, Yamaha created the 900 Seca from their comprehensive catalog of existing hardware, and threw in some new goodies for good measure. Their previous powerhouse, the XS1100, was handily outperformed by the new 900 Seca. Designed and built for a variety of riding styles, the 900 did almost everything well. The 853cc motor produced 85 horsepower, and returned over 50 miles per gallon in fuel mileage. Combined with the long wheelbase, comfortable seat, and stable handling, the 900 Seca was a near perfect sport/touring mount. The quality of the big Seca was overshadowed by several competing machines, and much to Yamaha's displeasure, it never sold in huge numbers. Those who own them still put them to good use today.

The actual dimensions of the Seca 900 were cloaked in a sleek cover of sheet metal and plastic, giving it the look of a much smaller model.

Slowing the Seca down from terminal velocity was completed by applying pressure to the twin, 10.9 inch discs on the front wheel.

Top Center:
The big, in-line displaced 853 cc's and came well dressed with an additional oil cooler to keep things under control.

One of the features that made the Seca such a great touring mount also kept it from being a true sport bike. The shaft drive added smoothness but used up horsepower.

Left:
Additional stopping power was available through the use of the 10.9 inch disc at the rear.

113

No sport/touring model would be complete without a comfortable place to sit. The 900 was gifted with a well shaped, nicely padded saddle for long days on the road.

With all of the necessary components tucked neatly within the frame, the Seca had a very narrow profile, which added to the confident handling of the bike.

In case you had the urge to carry a passenger along, there was a nicely finished grab rail within easy reach.

From nearly any angle, the 900's sleek lines convinced the eye that it was seeing a small sport bike, not a large sport/touring machine.

1984 Honda Nighthawk S

Owner: Steve Searles

In the very early part of the 1980s, Harley-Davidson began squealing in protest over the unfair competition coming from overseas. They alleged that the 750cc and larger machines from Japan were cutting into their market share. For the most part, these were two entirely different crowds, but the government responded by slapping a tariff on bikes over 750cc's. Since the brunt of sales were in the 750 range, the Japanese responded by building the tariff beaters. These were terrific machines that displaced no more than 700cc's, but still packed a sizeable wallop. At this point, the only machines affected by the restriction were the big liter bikes, and they had never made up that large a percentage in sales anyhow. The tariff was soon relieved after Harley began racking up huge sales gains of their own volition, and no longer sought protection to survive.

The 700S was built in response to a new government tariff, creating a new standard for the Universal Japanese Motorcycle (UJM) in the process.

Left: Although considered an odd choice for a sport bike, the 700S series were given shafts for their final drive.

Left:
The minuscule handlebar mounted fairing did not do much to smooth airflow, but managed to keep a few insects off of the rider's chest.

Far Left:
The 700S featured an all new, 696cc motor, complete with a hydraulic valve train. No more floating valves at high rpm and quieter performance were the by-products of the new design.

A frame mounted oil cooler was on duty to assist in keeping temperatures under control during spirited riding sessions.

The tail section found itself paired with an angular tail light and an aluminum grab handle.

The rear shocks had adjustment for both pre-load and damping. They did an admirable job of keeping the rider comfortable on all day jaunts.

The four, 31mm carbs, as well as the generator, were tucked tightly within the engine bay to keep the 700S narrow.

The dual discs were grooved for better rain management and were grabbed by two-piston calipers.

The 700S was offered in two color schemes for 1984. One would be hard pressed to choose between the two.

1984 Kawasaki 900 Ninja

Owner: Keith Olliges

Ever since the first Z-1 hit the streets of America in 1973, Kawasaki had begun to forge their reputation as the performance bike company. The other manufacturers built machines for speed as well, but Kawasaki always seemed to get a leg up on the competition with each new version of the hardware. The 900 Ninja combined the brute power of a water cooled 900cc motor with the style of a well dressed cafe bike, and created a whole new genre of sportbike. Capable of leaping tall drag strips in a single bound, or gobbling up miles of interstate, the 900 was equal to either task.

The Ninja warrior was considered the fiercest in all of Japan. This Kawasaki did its level best to live up to the name.

The aggressively styled fairing did more than look great, it actually helped keep the rider in a zone of clean air.

Tucked discreetly within the front of the fairing was a large, curved radiator to provide the Ninja with adequate cooling.

The front forks were fitted with AVDS, or Automatic Vehicle Damping System, a method of keeping the front end from diving under heavy braking.

The fuel filler cap looks like it was borrowed from a stealth fighter jet and was much neater than the earlier caps requiring a space when they were removed.

Another disc brake was mounted to the rear wheel and was used at times to slow the Ninja from ultra-legal speeds.

The entire exhaust was finished in a gloss black coating that looked a lot like the blueing of a gun, probably due to the fact that the spent gases left at about the same rate as a bullet.

Nestled within the steel chassis and plastic bodywork was a potent 900cc motor that launched the 900 Ninja into history as one of the top performing machines of all time.

The Ninja was beautifully finished, featuring these cast aluminum foot peg brackets which were strong, light, and terrific to look at.

The no nonsense office of the 900 rider included cast handlebars and three basic instruments, all without the clutter of idiot lights to distract him during a launch.

1985 Kawasaki Eliminator

Owner: Ray Landy

The Eliminator was designed to fill the needs of America's everyday riders. Being low, fast, and easy to maintain, the Eliminator filled the niche of "cruiser" for Kawasaki quite nicely. Powered by an altered Ninja motor of 908cc's, the Eliminator offered more torque, but less horsepower, than its sportbike sibling. A larger radiator and a totally different exhaust were two of the other modifications made to the Ninja motor being installed in the Eliminator. The upright seating of the Eliminator provided comfort whether cruising the downtown streets or hitting the open road. The four, 32mm carburetors provided sharp response to a twist of the throttle, and the Eliminator was one of the quickest machines in the cruiser field.

Kawasaki entered the Eliminator into the "cruiser" class for 1985, setting their sights on bringing performance and style together.

A pair of drilled disc brakes provided the stopping power up front.

Another departure from the Ninja was the shaft providing the final drive to the rear wheel.

No, it's not Las Vegas, just an amazing set of warning lights set between the chrome bezels of the instruments.

Looking a lot like the bill of a Platypus, the rear fender was fitted with this splash guard to keep the rider's back unstriped.

The fuel cap resided in a flip up housing; the short, narrow handlebars provided a comfortable reach for a majority of riders.

The radiator on the Eliminator was larger than the Ninja's and hung right out in front, instead of hiding behind a full fairing.

The performance was provided by an altered 908cc motor taken from the sporty Ninja.

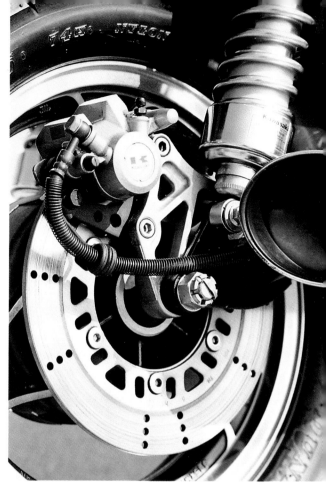

The huge rear tire was also given its own disc brake to assist in the slowing department.

The newly tweaked Eliminator motor exhaled through an entirely different set of exhaust pipes than that of the Ninja.

1985 Yamaha V Max

Owner: Matt Kallas

The history books show us that there is usually one kid in every class that makes it tough for the balance of the students. When the V Max first exploded onto the scene, the class was not only in trouble, but ran to hide. Boasting 145 horsepower and rakish styling, the V Max was like nothing ever seen before. There was no way to excuse the Max from the superbike class, so it went to the top in short order. With a motor pulled from the Venture Royale, then massaged to perform, the V Max set the performance bar way over the heads of every street bike available. The V Max is still in production today, and little has changed. The big V is still the King of the superbike hill, and proud of it.

From whatever angle you choose to view the V Max, you will never be bored by the circus of shapes or the plethora of hardware.

While the rider was given a sternly raked seat to hold him in place, the passenger was retained via the padded backrest.

The white faced instruments add an air of innocence to this bar room brawler but do little to convince you of an innocent nature.

Once in awhile the V Max pilot would have to slow down, so the machine was fitted with a trio of ventilated disc brakes.

Bottom Left: The tail end of the V Max is what challengers saw the most, they admired the deeply flipped fender as you rocketed past.

Bottom Right: The exhaust note of the V Max ranged from a syncopated rumble at idle to the well tuned shriek of a small block V8 at full chat.

The heart of the speed demon was a hot-rodded V four from the Venture Royale touring rig. There has never been another bike to match the numbers the V Max provided.

The tiny headlight seems woefully inadequate since the V Max could run near the speed of light itself.

Sure they're fake, and only for style, but no one would question your sanity with 145 horses on tap.

The V Max made no excuses for its fake air intakes or radical posture, it didn't need to. Any arguments could be easily sorted out at the drag strip where the V Max was capable of returning E.T.'s of under 11 seconds in street trim.

1986 Honda V65 Magna

Owner: Rick Youngblood

The prolific Honda catalogs have offered a variety of "cruiser" styles throughout the years. One of the most powerful was the V65 Magna. The throbbing V four engine delivered a high smiles-per-mile rating, and had a torque band that seemed to go on forever. With its raked front end, there was even a hint of the earlier "chopper" in its blood. With the big motor being water cooled, the Magna could retain a calm demeanor regardless of the riding conditions it was subjected to. The whole genre of cruisers was beginning to experience a slow down in sales, but would soon be resurrected and grow into one of the strongest selling "flavors" available.

By combining the virtues of V-four power, water cooling, and shaft drive, the Magna was touted as the ultimate in "cruisers."

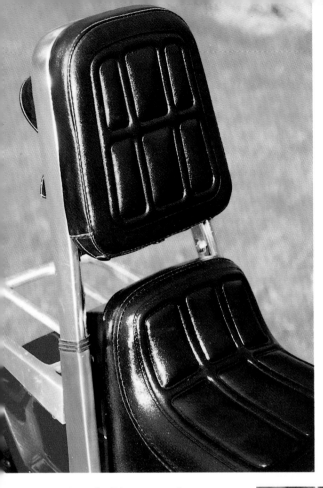

Not only did passengers have a comfortable spot for their derrieres, they were also given a thickly padded backrest for added pleasure.

Top Center:
The compact design of the Magna's motor provided a low center of gravity and decent handling for a bike with a raked front end.

The healthy radiator was bolted to the frame and kept the big four cool in the hottest of conditions.

In an effort to keep the rakish front end in line, a sturdy fork brace was included in the Magna's bag of tricks.

More than adequate braking was aboard, including a disc brake on the aft wheel.

The fully adjustable shocks gave the rider and passenger plenty of options to custom tailor their Magna to their needs.

The main braking force was provided by a pair of discs on the front wheel.

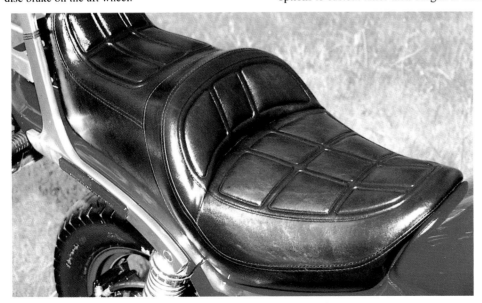
The deeply sculpted perch allowed for long days on the open road, even the passenger's section was accommodating.

The mild pullback of the bars gave the rider both a comfortable riding posture and an opportunity to utilize the bank of instruments.

1986 Suzuki GSX-R 750

Owner: Martin Horn/Aero P.A.

When designing the all new GSX-R, Suzuki engineers set out to create a racing bike for the street. More than simply a smattering of decals on a common cycle, they wanted their new entry to set new levels of lightness and performance. With a target weight of 388 pounds, and a dream of over 100 hp, they certainly had their work cut out for them. The end result was extremely close to their goals, and nothing short of spectacular. The GSX-R was the closest thing to a street legal racebike the world had ever known. From the dual headlights to the solo seating arrangement, the GSX made the lowliest of riders feel, and look like Kenny Roberts when astride the new Suzuki. Constant revisions continue today, and the GSX-R 750 is still the crown jewel in the pure sportbike field.

The dual headlights could be found on an endurance racer. Suzuki had just tucked the 24 Hours of LeMans trophy under their belt before the GSX hit the street.

Except for the headlights and turn signals, the new GSX-R looks every bit the street ready race bike.

The aluminum alloy frame was computer designed, offering all the strength of steel without the weight penalty.

Top Center:
The Full Floater Suspension was revised and lightened for application on the GSX-R. Weighing less than the previous version, the Full Floater still supplied the rider with a wide range of comfort and performance settings.

The dual discs mounted to the front wheel measure 11.8 inches in diameter and are grabbed by opposing calipers.

The rear set controls, even the footpegs, were honed from lightweight aluminum alloy to save a few precious ounces.

The fuel tank held 4.2 gallons of liquid and gave the look of an animal on its haunches, ready to pounce.

The narrow tail piece and solo seating do nothing to dispel the imagery of a racing machine.

On the rare occasion you may want a passenger aboard, the rear section lifted off to reveal another segment of seat.

The layout and accommodations were austere, but included everything the modern racer ... er ... rider would need.

1986 Suzuki RG500 Gamma

Owner: Rex Barrett

Since the RG500 was never formally imported into the U.S., it is not a common sight on the highways and byways. Considering the performance on tap, it is a shame they were not more readily available to sport riders in this country. Weighing only 340 pounds makes the RG one of the lightest machines around. Powered by a square four, two stroke motor, the Gamma delivers a high octane punch at higher rpm's that can be matched by no four stroke machine invented. With 90 horses available at the twist of your wrist, the RG is not for the feint of heart. It may be ridden around town in complete harmony, and can turn into a snarling beast on the open road by spinning the tach in the upper reaches of the 13,000 rpm redline. With a delivery of power much like a light switch, the Gamma can be quite a handful to the untrained rider.

Built as a thinly disguised street legal race bike with two stroke power, the RG500 Gamma was a bit of an enigma, answering a question that no one had really asked. The Yamaha RZ500 was its only competition and both were phased out soon after production.

Within the confines of the aluminum box-section frame is a square four motor that produces 90 horsepower using two stroke technology.

Powerful brakes were available on demand and were fitted with an anti-dive compensator to keep things under control while stopping the RG.

One pair of the exhaust outlets ran high on the chassis and were kept cool through extensive venting of the bodywork.

Suzuki put their full floater suspension to work on the Gamma in an effort to continue the race-bike-for-the-street theme.

Both the fuel tank and the aluminum chassis are marked with the Greek "Gamma" symbol.

The high mounted exhausts made their exit just below the tail light, while a second pair took a more conventional route.

Although a bit cramped, the solo seat is actually quite comfortable. It even has a tiny pillion for a passenger hidden beneath the fiberglass cover.

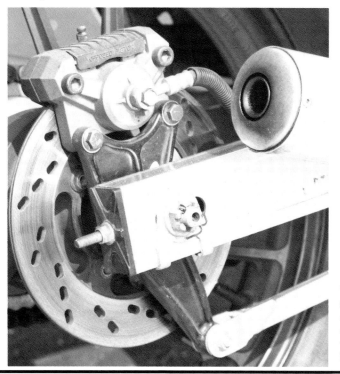
The box section swingarm provided extreme rigidity and was augmented with a third disc brake.

1989 Honda GB500 Tourist Trophy

Owner: Dennis French

The usual indication of a classic design is its enduring beauty. Through thick and thin, good times and bad, a classic machine always comes up looking as good as the day it was introduced. The design team at Honda chose the Brit bikes of yesteryear to use as their inspiration for the GB500. Starting with the parallel twin motor, and leading back to the sweep of the tail piece, the Tourist Trophy looks every bit the British machine. The elegant, deep green paint, trimmed with only a simple gold pinstripe, evokes images of a heavy fog hanging over cobblestone streets. The more you stare at the GB, the more settled it appears. Every line seems to belong just where it lays. It looks like a classic.

The simple design of the early British machines inspired the designers at Honda to create a modern day classic of their own.

The sidecovers are emblazoned with a regal crest as well as the "Tourist Trophy" moniker.

From any angle, the Tourist Trophy drips elegance, not oil, which breaks with the tradition of typical British two wheelers.

Prepare to spend a lot of time alone when riding the GB, because beneath the rear cover is not a seat, but a small storage area.

Stopping power is supplied via a single disc brake up front, providing plenty of grip on a light weight machine like the GB.

The footpegs and exhaust are supported by this Gibraltar looking casting; however, the casting is quite light due to the use of aluminum alloy in place of steel.

The twin cylinder motor produces adequate power, yet remains compact and simple.

The singular, seven inch headlight is topped by a pair of chromed instrument pods to complete the distinctly British Tourist Trophy.

The suspension on the front fork is gifted with the gaiters always found on machines hailing from across the pond.

Replacing the standard tubular steel bars are these short, cast aluminum units that allow some adjustment for the rider's reach.

1990 Honda RC-30

Owner: Matt Jonas

When Honda designed the RC-30, they had a lot more in mind than simply another full faring, racer wanna be. It certainly has the look of a full fledged race machine, but the hot rod heart of the RC can be found by digging beneath the exterior trim. Once inside the 750cc double overhead cam, V four motor, you will find titanium connecting rods. Not your average street going fare to be sure. The front wheel has been fitted with hardware that allows for lightning fast changes of both the wheel and the brake pads. Street bike indeed. Out back you discover a single sided swing arm that holds the rear wheel in place with a single lug nut. Is it a street bike ... or a cleverly disguised race bike? Riding the RC would point one's answer toward the track, but then Honda has built their reputation by producing some of the raciest bikes in the world.

Thinly disguised as a street bike, the RC-30 is only inches away from being race ready.

The sculpted fuel tank was topped off with a flush fitting, aircraft style door and came complete with a race type vent tube.

Other subtleties separating the RC-30 from the track were a pair of halogen headlights and turn signals.

Don't even think of bringing a passenger along for the ride, for there is no seating available for a second participant.

All four exhaust tubes led into this polished canister on the drive side of the chassis. Placing it here kept the access to the quick change wheel open for quicker installation of fresh rubber.

The fact that this machine even wears a license plate is a bit foolish, since it is difficult to consider it anything other than a race bike for the street.

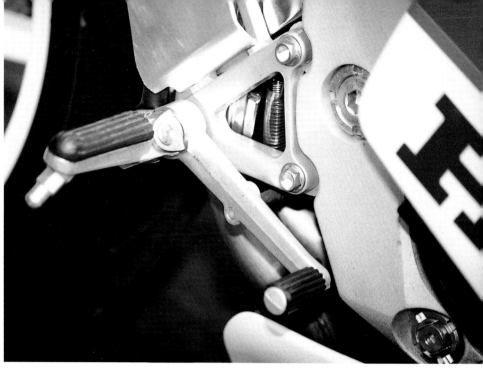

The aluminum alloy rear sets are but a few of the lightweight components you can find on the RC.

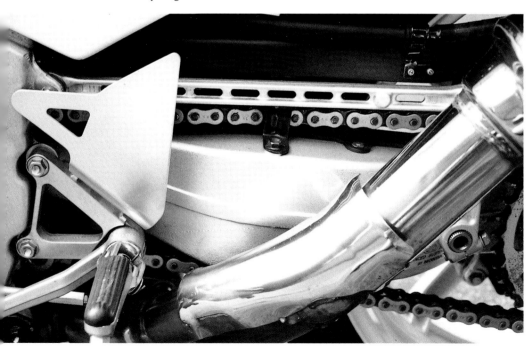

The single sided swingarm was massive, providing enough stiffness to keep the RC firmly planted during high speed cornering.

The full bank of instruments are certainly an indication of the RC-30's street going accommodations.

The spidery web of the rear wheel can be easily removed by loosening the single, albeit large, lug nut.

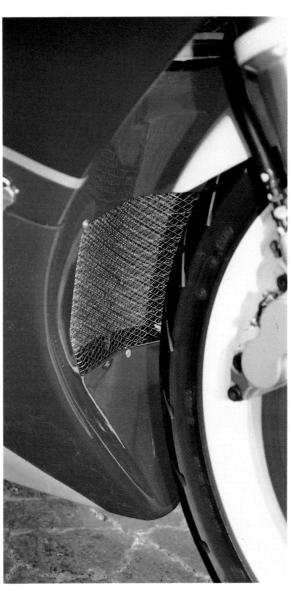
Inside the full coverage fairing lies a radiator to keep the 750cc's of power cool during the heat of battle.

Another piece in the racebike/streetbike mystery was the use of the gigantic front disc brakes that could easily be swapped.

1996 Honda CBR600 F3

Owner: Kevin Anderson

The CBR600 from Honda has repeatedly proven itself to be a competent cycle both on the sales floor and on the race track. The nimble handling and knife edge performance make it a popular choice for sport minded riders everywhere, including on the track. The CBR won the Daytona 200 in 1995, with Miguel Duhamel aboard. To commemorate their victory, Honda introduced the Smokin' Joe's replica for street bound riders. Finished in the distinctive purple and yellow livery of the race bike, this limited edition is tough to mistake for any other scoot in the parking lot. In addition to the striking paint scheme, the replica racer is given a full brace of racing sponsors decals. None of this finery takes away from the superior handling characteristics of the F3; it just makes it look fast while sitting still.

The two tone paint scheme and racing graphics make the replica model look every bit the real race bike.

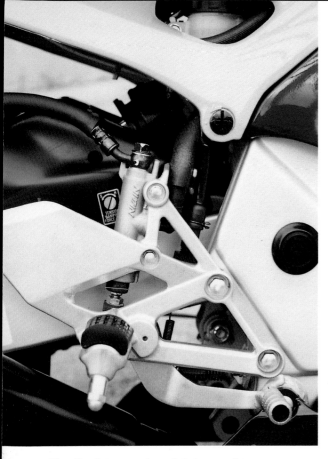

The alloy foot controls, and their mountings, are geometric works of art that provide strength and beauty at the same time.

Hidden within the confines of the full fairing is the 600cc, water cooled motor that carried the F3 into the winner's circle at racing venues all around the United States.

Radial tires, alloy rims, and floating brakes provide the illusion of a race machine for the street.

The remote brake reservoir appears to be taken right off the race bike and provides exceptional stopping power to the rear disc.

The racing trim extends all the way to the box section swingarm, adding to the feel of a true race bike.

The racing trim starts on the front fender and fork lowers, and continues all along the length of the F3.

With aerodynamics playing an important role in a race bike's design, the CBR features a flush mounted headlight in lieu of the number plate of the racing version.

About the only non-racing piece on the F3 is the two place seat. A real racing bike has room for the pilot, and the pilot alone.

Adding an air of authenticity, each Smokin' Joe's replica comes complete with Miguel Duhamel's signature on the tank.

1996 Honda VFR 750

Owner: Wil Mateo

In a motorcycle world filled with narrowly focused machines to satisfy every rider's fantasy, it is refreshing to know there is at least one machine capable of doing it all. The VFR 750 has garnered "Best 750" awards from several major cycle magazines in the past year, and for good reason. The bright red VFR can tool around town without raising an eyebrow, or soak up huge lengths of highway at hair-raising speeds. The V-four motor has a seamless power band, and pulls right from idle, all the way to the redline without a hitch. It can carry two people in comfort, or provide a solo rider with his or her own nirvana. The fairing and windshield provide a comfort zone at speed, and gives the VFR a distinctive race bike look. About the only drawback to the VFR is its flaming red finish. There is no color that raises the hackles of the police faster than "Arrest me red."

Despite its look of a pure racing machine, the VFR 750 is a motorcycle capable of providing thrills, no matter what style of riding you prefer.

Left: Behind the cheese grater vent lurks the potent V four motor. The black N.A.R.T. duct only adds to sporting nature of the VFR's appearance.

Right:
The five spoke wheel mounts to the single sided swingarm with four lugs and is slowed by a single disc brake.

Left:
A pair of fully floating discs are grabbed by dual piston calipers, providing copious quantities of slowing power.

Below:
The disc brake, drive sprocket, and single sided swingarm all reside on the left side of the cast wheel.

Below:
With the twist of two retainers, the VFR is now the perfect one-man machine, and looks more like a race bike for the street.

The cat-like headlights are often the last thing seen before the VFR goes past in a flash of red.

The instrument panel of the VFR puts many automotive dashes to shame. The layout and completeness give the rider all his information in an easy to read configuration.

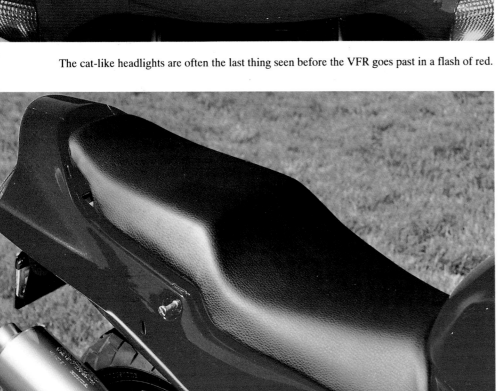

Set up for carrying two people in comfort, the VFR proves it is more than a solo machine.

Unblemished by gaudy graphics, the fuel tank is a field of pure red — which does nothing to help the VFR hide in the shadows.

With the exposed frame rail and sleek bodywork, the VFR is not an easy bike to miss as it cruises past at hyper-legal speeds.